Under the
Microscope:
Backyard Bugs

LADYBUGS

PowerKiDS
press
New York

Suzanne Slade

To Gloria

Published in 2008 by The Rosen Publishing Group, Inc.
29 East 21st Street, New York, NY 10010

First Edition

Editor: Joanne Randolph
Book Design: Julio Gil
Photo Researcher: Nicole Pristash

Photo Credits: Cover, pp. 1, 7, 9, 11, 13, 15, 17, 19, 21 © Shutterstock.com; p. 5 © istockphoto.com/Sven Brandt; pp. 11 (inset), 15 (inset) © Dennis Kunkel Microscopy, Inc.

Library of Congress Cataloging-in-Publication Data

Slade, Suzanne.
   Ladybugs / Suzanne Slade. — 1st ed.
      p. cm. – (Under the microscope: backyard bugs)
   Includes index.
   ISBN-13: 978-1-4042-3818-3 (lib. bdg.)
   ISBN-10: 1-4042-3818-2 (lib. bdg.)
   1. Ladybugs—Juvenile literature. I. Title.
   QL596.C65S53 2008
   595.76'9—dc22
                                        2006036969

Manufactured in the United States of America

Contents

Right in Your Own Backyard

Do you know what a **habitat** is? A habitat is any place that animals and plants call home. Even your backyard is a habitat!

When warm weather arrives in spring, the habitat in your backyard comes alive. Brown grass turns a bright shade of green. Leaves appear on bushes and trees. Colorful flowers push up from the ground. Many kinds of wildlife come out of hiding. Some bugs slowly crawl across the ground. Those with wings take flight in the air. Look closely and you might see a tiny red bug with black spots zoom by. You just discovered a ladybug!

Many people say that if a ladybug lands on you, you will have good luck. This likely comes from the fact that ladybugs are helpful bugs.

Meet the Ladybug

The ladybug belongs to a group of **insects** called beetles. A beetle has two wings that close over its back to form a hard outer shell. These wings are called forewings. A beetle also has a pair of soft wings that fold inside its shell. Some people call the ladybug a ladybird beetle or lady beetle.

Ladybugs are an important part of your backyard habitat. You can find this tiny insect flying through the air, crawling on a swing, or sitting on a plant. Ladybugs eat tiny bugs that feed on flowers, trees, and other plants. Larger bugs eat ladybugs for their food.

Most people recognize a ladybug because of its spotted shell. This colorful shell is actually the ladybug's forewings.

Lots of Ladybugs

There are more than 4,000 **species**, or kinds, of ladybugs in the world. In North America you can find more than 350 different ladybug species. Red ladybugs with black spots are the most common in the United States. Many species have a yellow or orange outer shell with black spots, though. Some ladybugs are black with red spots. Others have no spots at all.

Many ladybugs get their names based on color and markings. The seven-spotted ladybug has seven black spots on its red shell. A three-banded ladybug has three black stripes across an orange shell.

You may think of a ladybug as a red bug with black spots. There are lots of other kinds of ladybugs, though, like this black one with orange spots.

Ladybug Parts

Like all insects, a ladybug's body is made of three main parts. The first part is its small, black head. A ladybug has two **antennae** on its head. Ladybugs use their antennae to touch, taste, and smell.

The middle part of the ladybug is called the **thorax**. The hard forewings keep the thorax safe. The ladybug's other pair of wings is tucked between the thorax and forewings. Ladybugs also have six legs, for climbing and walking. The **abdomen**, or stomach, is the third part of a ladybug. This is where a ladybug breaks down its food.

Both sets of wings can be seen on this flying ladybug, as well as the abdomen, the thorax, and the head. *Inset:* This is a close-up photo of a ladybug's head.

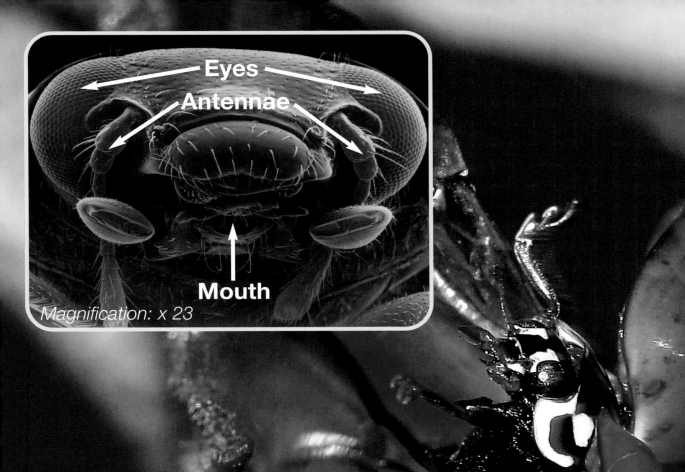

Eyes

Antennae

Mouth

Magnification: x 23

Laying Eggs

Not all ladybugs are ladies, or female. There are plenty of male ladybugs, too. In the warmth of spring and summer, female and male ladybugs **mate**. About one week later, the female lays her eggs.

She will search for safe places with plenty of food to put the eggs. Ladybugs often hide their eggs on the bottom side of a leaf. The eggs stick where they are placed. The female leaves about 25 to 50 eggs in each spot. She will lay about 200 eggs total. The tiny yellow eggs **hatch** about eight days later.

After mating with a male ladybug, a female can sometimes wait two or three months before she lays eggs.

13

Growing Up

A ladybug begins its life as an egg. The egg turns gray when it is ready to hatch. Soon a **larva**, with its dark body and six legs, wiggles out. A larva eats bugs called aphids. It can eat up to 150 aphids in one day!

As it eats and grows, a larva **sheds** its old skin. The larva eats for three weeks and sheds four times. Then it glues itself to a plant and turns into a **pupa**. Seven days later, the pupa becomes an adult ladybug.

A ladybug spends about 12 days as a larva. It grows to be 12 times bigger than when it started out. *Inset:* This is a photo of a ladybug larva taken with an electron microscope.

Magnification: x 4

15

The Helpful Ladybug

Just like larvae, adult ladybugs eat aphids. Aphids are tiny bugs that suck sap out of plant stems. These small plant eaters can destroy a whole field of crops. Adult ladybugs eat about 100 aphids in a day.

Some farmers use ladybugs to keep their plants safe from pests, instead of using costly or unsafe sprays. They buy a large number of ladybugs to put in their fields to get rid of the hungry aphids. Another type of ladybug, the Australian beetle, eats a different farm pest. These black ladybugs eat mealy bugs, which also destroy crops.

These ladybugs are eating black aphids. Black aphids do a lot of harm to plants and are hard to kill with sprays.

Ladybug Enemies

The ladybug has very few enemies. Large bugs, such as the shieldbug, assassin bug, and praying mantis, eat ladybugs. If a ladybug walks into a web, it might become a spider's dinner. Sometimes people hurt this helpful insect with bug spray.

A ladybug knows some tricks to stop its enemies. If a bird or small animal is near, it can play dead. The ladybug can also give off a bad smell, which scares animals away. When a ladybug is afraid, its legs produce a bad-tasting juice. This makes some animals think twice before eating the ladybug.

This praying mantis is on the lookout for food. The mantis uses its long front legs to catch its next meal.

Winter Rest

Before winter arrives, most ladybugs fly to warm places to rest. This rest is called **hibernation**. Like many insects, ladybugs cannot live in cold weather. Some ladybugs hibernate together in large groups. They look for places away from the wind. Ladybugs may hide in openings or cracks in tree trunks, under rocks, or inside buildings.

During winter, plants die and there are no aphids to eat. Ladybugs store up fat and sugar in their body, so they can last five months without food. The warmth of spring wakes ladybugs up from their long winter rest.

Most ladybugs live about one to two years. When a large number of ladybugs are found together in one place, it is called a swarm.

Ladybug Hunt

Now that you know more about ladybugs, why not go on a ladybug hunt? On hot sunny days, look for ladybugs on plants or in the air. They can fly fast with wings that beat 85 times each second.

Search for oval-shaped eggs and larvae on plants. A young larva is short and black, while an older larva is grayish yellow and about 1 inch (2.5 cm) long. Can you find an orange and black pupa glued to a leaf? Watch this tiny ball closely, and soon you can welcome a brand new ladybug to your backyard habitat!

Glossary

abdomen (AB-duh-mun) The large, back part of an insect's body.

antennae (an-TEH-nee) Thin, rodlike feelers on the head of certain animals.

habitat (HA-beh-tat) The kinds of land where an animal or a plant naturally lives.

hatch (HACH) To come out of an egg.

hibernation (hy-bur-NAY-shun) Spending the winter in a sleeplike state, with heart rate and breathing rate slowed down.

insects (IN-sekts) Small animals that often have six legs and wings.

larva (LAHR-vuh) An insect in the early life period in which it has a wormlike form.

mate (MAYT) To join together to make babies.

pupa (PYOO-puh) The second stage of life for an insect, in which it changes from a larva to an adult.

sheds (SHEDZ) Gets rid of an outside covering, like skin.

species (SPEE-sheez) One kind of living thing. All people are one species.

thorax (THOR-aks) The middle part of the body of an insect. The wings and legs attach to the thorax.

Index

A
abdomen, 10
antennae, 10
aphids, 14, 16, 20
assassin bug, 18
Australian beetle, 16

C
crops, 16

F
forewings, 6, 10

H
habitat, 4, 6, 22

I
insect(s), 6, 10, 18

L
larva(e), 14, 16, 22

M
mealy bugs, 16

N
North America, 8

P
plant(s), 4, 6, 14, 16,
    20, 22

praying mantis, 18
pupa, 14, 22

S
sap, 16
shell, 6, 8
shieldbug, 18
species, 8

T
thorax, 10

Web Sites

Due to the changing nature of Internet links, PowerKids Press has developed an online list of Web sites related to the subject of this book. This site is updated regularly. Please use this link to access the list:
www.powerkidslinks.com/umbb/lbug/